Sabrina Mejdoub
Lalla Mariem Cheikh Hamza

Anti-centromere antibodies in routine use

Sabrina Mejdoub
Lalla Mariem Cheikh Hamza

Anti-centromere antibodies in routine use

Technical aspects and clinical relevance

ScienciaScripts

Imprint

Any brand names and product names mentioned in this book are subject to trademark, brand or patent protection and are trademarks or registered trademarks of their respective holders. The use of brand names, product names, common names, trade names, product descriptions etc. even without a particular marking in this work is in no way to be construed to mean that such names may be regarded as unrestricted in respect of trademark and brand protection legislation and could thus be used by anyone.

Cover image: www.ingimage.com

This book is a translation from the original published under ISBN 978-620-3-45875-6.

Publisher:
Sciencia Scripts
is a trademark of
Dodo Books Indian Ocean Ltd. and OmniScriptum S.R.L publishing group

120 High Road, East Finchley, London, N2 9ED, United Kingdom
Str. Armeneasca 28/1, office 1, Chisinau MD-2012, Republic of Moldova, Europe
Managing Directors: Ieva Konstantinova, Victoria Ursu
info@omniscriptum.com

Printed at: see last page
ISBN: 978-620-3-50029-5

ANTI-CENTROMERE ANTIBODIES IN ROUTINE USE :

TECHNICAL ASPECTS AND CLINICAL RELEVANCE

DR SABRINA MEJDOUB

DR LALLA MARIEM CHEIKH HAMZA

PLAN

INTRODUCTION

Anti-nuclear antibodies (ANA) are the autoantibodies most frequently prescribed in clinical practice. These antibodies are directed against different antigenic targets in the cell nucleus (1). Testing for them is useful for diagnosing connectivites, which include: systemic lupus erythematosus (SLE), Gougerot-Sjögren's syndrome, systemic scleroderma (SSc), idiopathic inflammatory myopathies and mixed connectivitis (2).

NAAs are traditionally detected in two stages: a first stage of screening by indirect immunofluorescence (IFI) on Hep-2 cells which, if positive, enables the titre of NAAs and the appearance of fluorescence to be determined, and a second stage of typing NAAs to identify the antigenic target(s) recognised by these Ac (3).

Among these NAAs, anti-centromere mAbs are directed, as their name suggests, against the area of the chromosome where, during mitosis, the two sister chromosomes remain attached before separating (4). More specifically, anti-centromere antibodies recognise various proteins in the kinetochore, a structure that enables chromosomes to be attached to the fibres of the mitotic spindle, allowing them to migrate towards the two poles of the cell (4).

These anti-centromeric Ac can be identified by IFI on Hep-2 cells using a specific fluorescence pattern, but also by other techniques using centromeric proteins (CENPs) as target antigens (Ag) (4)(5).

With regard to their clinical value, anti-centromere antibodies have been reported to be specific markers of SSc; however, their specificity has been disputed by some authors (6).

The objectives of our study were: **1/** to evaluate the concordance of two immunological techniques used in the Immunology Laboratory of the Habib Bourguiba University Hospital in Sfax (Tunisia) for the detection of anti-centromere Ac, namely IFI on Hep-2 cells and immunodot, **2/** and to determine the clinical significance of the positivity of the anti-centromeric Ac detected by these techniques.

PATIENTS AND METHODS

1. PATIENTS

We conducted a retrospective study over a 2-year period (January 2020- December 2021). Among the requests for serological testing for NAAs sent to the Immunology Laboratory of the Habib Bourguiba University Hospital in Sfax, those that had undergone both IFI screening on Hep-2 cells and immunodot typing were identified. Patients with anti-centromeric Ac positivity by IFI and/or immunodot were included (Figure 1). For each patient included, we recorded the following data: gender, prescribing department, NAA titre, appearance of fluorescence and antigenic specificities identified by immunodot, as well as any clinical information available.

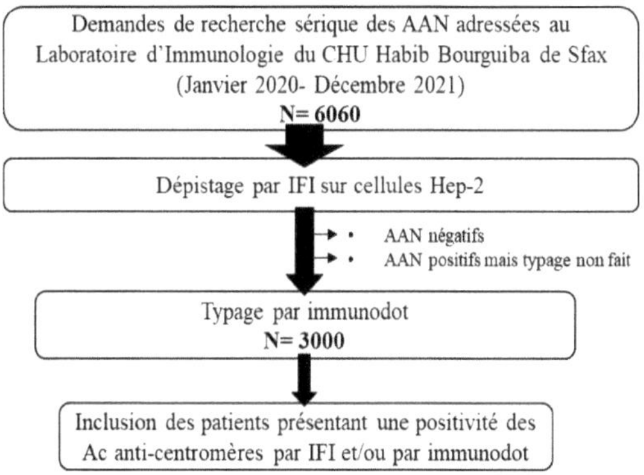

Figure 1: Recruitment strategy for the study population

2. METHODS

For each patient, a whole blood sample was taken in a dry tube and sent to the laboratory at room temperature. After clotting, the sample was centrifuged at 3,000 rpm for 10 minutes. The serum was decanted into a 5 ml haemolysis tube (stored at +4°C until analysis) and an eppendorf (stored at -20°C for reserve/serum bank).

2.1. Search for NAAs by IFI on Hep-2 cells

Serum detection of NAA (IgG isotype) was performed by IFI on Hep-2 cells using a commercially available kit "IIFT: HEp-2" (Euroimmun® , Germany).

2.1.1. Principle

The antigenic substrate fixed on the slides corresponds to Hep-2 "Human epithelioma cells" derived from a tumour line of human epithelial cells (laryngeal carcinoma). After incubating the slides with serum, any NAAs present in the serum bind to their antigenic targets. The Ag-Ac antigen reaction is then revealed by a secondary Ac (anti-IgG) labelled with a fluorochrome (fluorescein). This is read using a fluorescence microscope. (Figure 2)

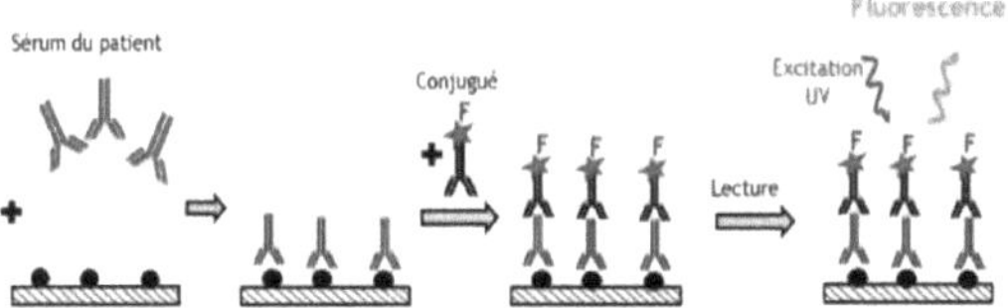

Figure 2: Principle of indirect immunofluorescence

2.1.2. **How it works**

The test was performed according to the supplier's instructions. The slides, conjugate, positive and negative controls and mounting medium were ready to use. To prepare the PBS-Tween buffer, one sachet of PBS is dissolved in 1 litre of distilled water to which 2 ml of Tween 20 is added, then shaken for 20 min until homogenised.

Patient sera were diluted in PBS-Tween buffer. The initial dilution was 1/160 for adults and 1/80 for children. In the event of a positive result, titration of Ac was carried out by cascade dilution to 1/2 (80 → 160 → 320 → 640 → 1280).

A volume of 30 µl of positive and negative controls as well as diluted samples were deposited in the reaction wells of the reagent support (TITERPLANE), avoiding air bubbles.

The slides were removed from their packaging and placed, face down, on the reagent support, ensuring that each previously deposited sample came into contact with its antigenic substrate.

After 30 minutes incubation at room temperature (+18°C to +25°C), the slides were rinsed and then immersed in a wash tank containing PBS-Tween under agitation for at least 5 minutes.

Next, 25 µl of the conjugate (fluorescein-labelled anti-IgG) was deposited on each reaction well of a clean reagent holder.

The slides were removed from the wash tank one by one, wiped (back and sides only) with a compress and then placed on the reagent support, checking that there was correct contact between the previously deposited conjugate and the antigenic substrate.

After a second 30-minute incubation at room temperature and in the dark, the slides were rinsed and then immersed in a newly-filled wash tray with PBS-Tween, adding a quantity of Evans Blue (150-200 µl), with shaking, for at least 5 minutes. Finally, coverslips (the number of slides used) were placed on the clean reagent support and drops of mounting medium were placed on the coverslips (drops of max. 10 µl per reaction well). The slides were removed from the wash tank, wiped clean and placed on the coverslips prepared in the reagent holder.

2.1.3. **Interpretation**

Readings were taken using a fluorescence microscope (40x objective). The NAA screening result was considered positive if nuclear fluorescence was observed with the initial dilution (1/160 for adults, 1/80 for children). The NAA titre corresponds to the inverse of the last dilution giving a positive result. Depending on the type of nuclear fluorescence observed, the appearance was specified (Figure 3):

- homogeneous: even fluorescence throughout the nucleoplasm. Nucleoli may or may not be labelled depending on the cell preparation. Cells in mitosis (metaphase, anaphase and telophase) have their chromatin intensely labelled in a homogeneous and hyaline manner.

- speckled: granular fluorescence of varying fineness and density. Nucleoli may or may not be labelled. The chromatin of cells in mitosis (metaphase, anaphase and telophase) may or may not be labelled.

- nucleolar: fluorescence of the nucleoli; the nucleoplasm can be negative (isolated nucleolar aspect) or fluorescent with a different fluorescence intensity (combination of two aspects)

- centromeric: In cells in interphase, around 40 large grains per cell are dispersed. In cells in mitosis, these grains are aligned and superimposed on the chromatin.

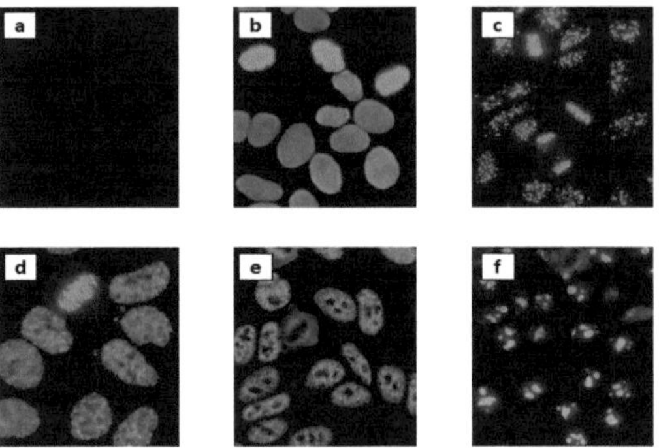

Figure 3: Nuclear fluorescence aspects

a: AAN negative; **b:** homogeneous appearance; **c:** centromeric

appearance; **d and e:** mottled appearance;

f: nucleolar aspect

2.2. Typing of NAAs by immunodot

NAAs were typed by immunodot (Kit "Euroline NAA Profile 3 plus

DFS70 (IgG)", Euroimmun® , Germany).

2.2.1. Principle

Antigenic targets bound to commercially available nitrocellulose

strips include: nRNP/Sm, Sm, SS-A, Ro-52, SS-B, Scl-70, PM-Scl

100, Jo-1, CENP-B (centromeric protein B), PCNA, native DNA, nucleosomes, histones, ribosomal protein P, AMA-M2 and DFS70 (Figure 4). After incubating the strips with the serum, any specific Ac present in the serum bind to their antigenic targets. The Ag-Ac reaction is then revealed by a secondary Ac (anti-IgG) labelled with an enzyme (alkaline phosphatase). The addition of the enzyme substrate allows, Following the enzymatic reaction, a coloured product is obtained which precipitates at the site of the Ag-Ac reaction. The result is a band whose colour intensity is proportional to the concentration of specific Ac. It can be read with the naked eye or with a scanner (Figure 5).

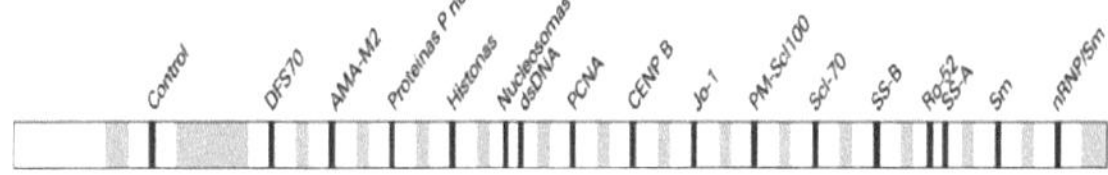

Figure 4: Strip used for typing NAAs by immunodot

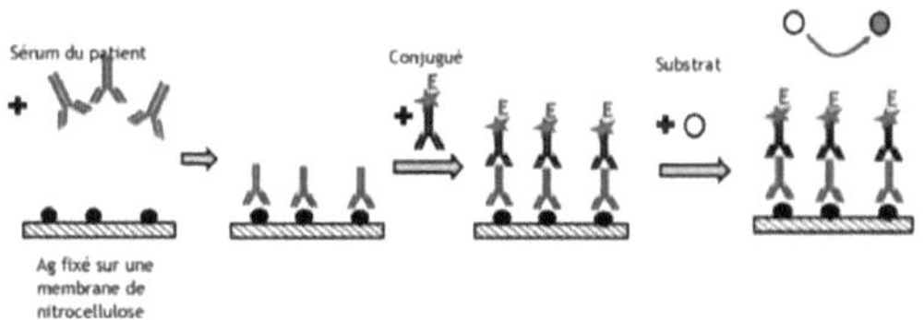

Figure 5: Principle of immunodot

2.2.2. How it works

The test was performed according to the supplier's instructions. After pre-treatment of the strips (incubation with 1.5 ml of dilution buffer under agitation for 5 min then aspiration), 1.5 ml of diluted sera (1:100 dilution with dilution buffer) were added to each incubation well and incubated at room temperature for 30 min with continuous agitation. This was followed by a wash step in which the diluted sera were aspirated from each well, 1.5 ml of wash buffer was added, shaken for 5 min and then aspirated. This procedure was repeated 3 times. Then 1.5 ml of conjugate (anti-IgG alkaline phosphatase) was added to each incubation well and incubated at room temperature for 30 min with continuous shaking. After the second washing step (identical to the first), 1.5 ml of enzyme substrate was added to each incubation well and incubated at room temperature for 10 min with continuous shaking. Finally, the reaction was stopped by aspirating the substrate and rinsing the strips 3 times with 1.5 ml of distilled water.

2.2.3. Interpretation

The scan was performed using the EUROLine Scan programme. The result is semi-quantitative (Figure 6).

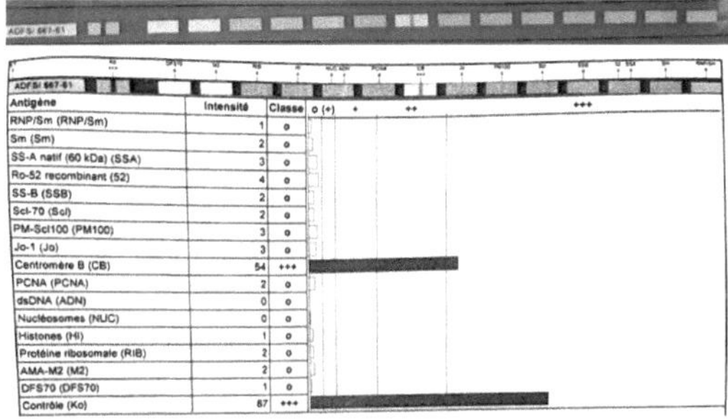

Figure 6: Example of an AAN typing result by immunodot (positive anti-centromeric Ac +++ and negativity of the other antigenic specificities tested)

2.3. Statistical study

Statistical analysis was carried out using Microsoft Excel. The Chi-squared test was used to compare the frequency of a qualitative variable between two groups. A p-value of less than 0.05 was considered significant.

RESULTS

During the study period, 63 cases of anti-centromere Ac positivity (by IFI and/or immunodot) were included, i.e. 2.1% of applications that had undergone both IFI screening on Hep-2 cells and immunodot typing.

1. CHARACTERISTICS DEMOGRAPHICS OF THE STUDY POPULATION

There was a clear predominance of women, with a male/female sex ratio of 0.12 (Figure 7).

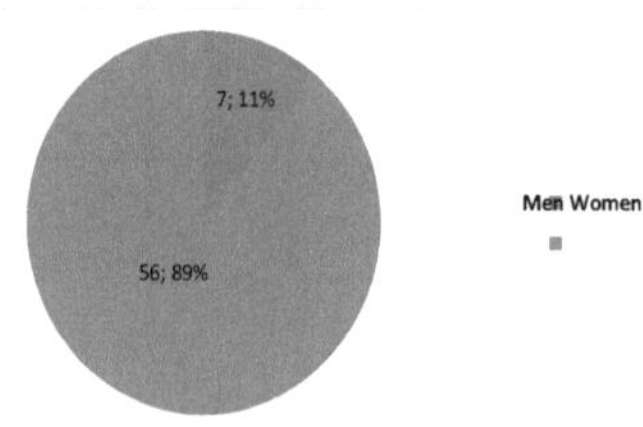

Figure 7: Breakdown of study population by gender

Age was specified for 18 applications. The average age was 39.7, with extremes ranging from 3 to 65.

2. BREAKDOWN OF CASES INCLUDED BY PRESCRIBING DEPARTMENT

The majority of the cases included were referred from the internal medicine department (31.7%). It should be noted that 9 cases were referred from a paediatric ward. Figure 8 shows the distribution of referrals by referring department.

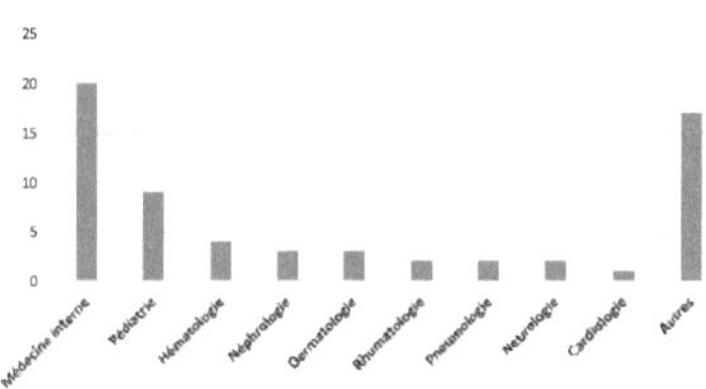

Figure 8: Breakdown of cases included by prescribing department

3. RESULTS OF AAN SCREENING

Anti-centromeric Ac was detected by IFI on Hep-2 cells in 20 cases (31.7%). In 3 cases/20, centromeric fluorescence was associated with speckled nuclear fluorescence. Other aspects of fluorescence observed were speckled (34 cases; 54%), nucleolar speckled (6 cases; 9.5%) or homogeneous (3 cases; 4.8%). The titre of the NAAs ranged from 1/160 (1.5%) to a titre greater than or equal to 1/1280 (61.9%). Figure

9 shows the distribution of included cases according to NAA titre and appearance.

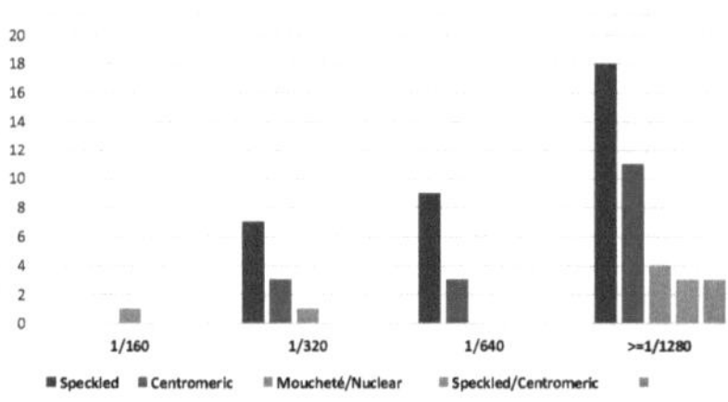

Figure 9: Breakdown of cases included according to title and aspect

of NAAs

4. AAN TYPING RESULTS

Anti-CENP-B centromere antibodies were detected by immunodot in 61 cases (96.8%). These antibodies were weakly positive (+: 29 cases), positive (++: 8 cases) or strongly positive (+++: 24 cases).

No specificity other than CENP-B was detected by immunodot in 30 cases. In the other cases, the most frequently associated specificities were anti-Ro-52 (n=10), anti-SS-A (n=7), anti-M2 (n=7) and anti-DFS70 (n=7). (Figures 10 and 11)

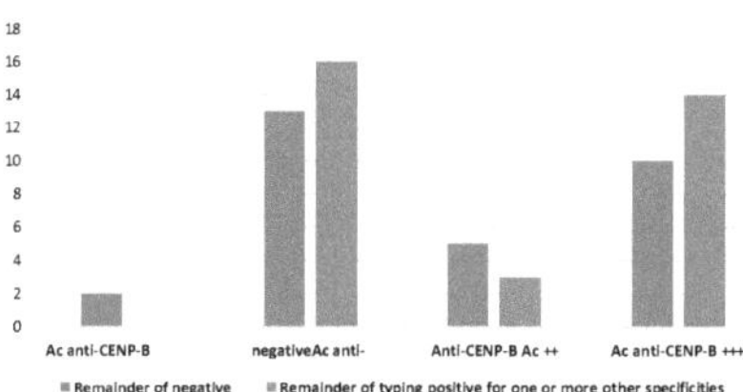

Figure 10: Distribution of cases included according to the result of

the anti-centromere (CENP-B) immunoassay.

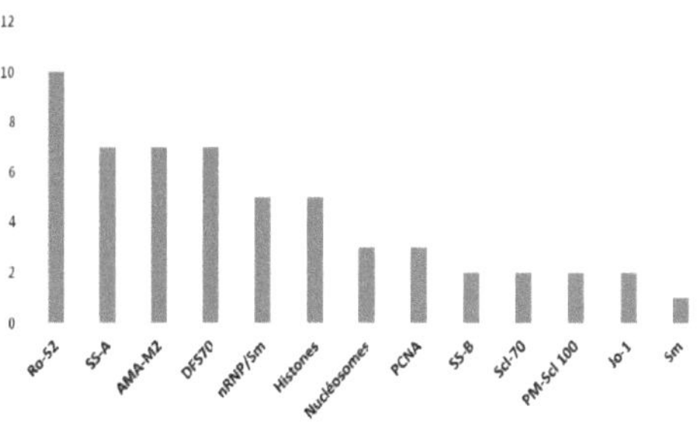

Figure 11: Frequency of antigen specificities (other than CENP-B)

detected by immunodot

5. CONFRONTATION OF RESULTS OF CA BY IFI AND IMMUNODOT

Figure 12 shows the distribution of cases included according to the

results of anti-centromere Ac by IFI and immunodot. The positivity of anti-centromere Ac by IFI on Hep-2 cells was confirmed by the positivity of anti-CENP-B Ac by immunodot in 18 cases. The discrepancies were 2 cases of anti-centromere Ac positivity by IFI and anti-CENP-B Ac negativity by immunodot, and 43 cases of anti-centromere Ac negativity by IFI and anti-CENP-B Ac positivity by immunodot.

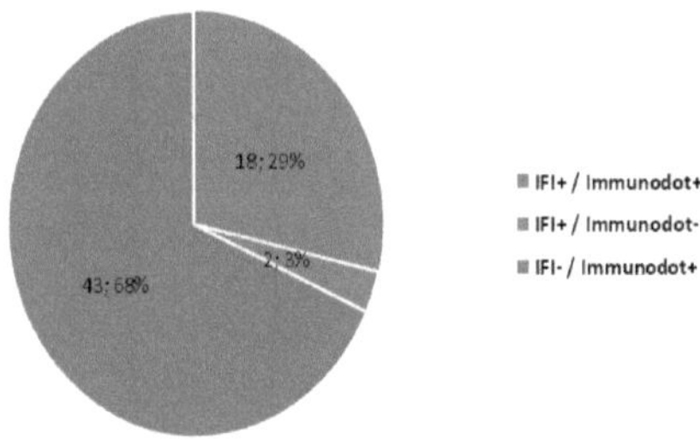

Figure 12: Distribution of the study population according to the result of anti-centromere Ac by IFI and immunodot

For the 2 IFI+ / Immunodot- cases, the centromeric appearance was observed in isolation with a titre of 1/320 in one case and 1/1280 in the other. Immunodot typing was negative for all antigenic specificities tested. Among the cases of anti-centromere Ac positivity by immunodot, we compared concordant cases (IFI+ / Immunodot+) and discordant cases (IFI- / Immunodot+) (Table I). Anti-CENP-B Ac positivity at 2 or 3 crosses was significantly more frequent in IFI+ cases (15/18; 83.3%) than in IFI- cases (17/43; 39.5%) (p=0.002). There was no significant difference between the two groups with regard to the association of anti-CENP-B Ac with other antigenic specificities on immunodot (55.6% versus 53.5%; p=0.883).

Table I: Comparison of results of NAA screening and typing in patients with anti-centromeric Ac detected by immunodot and IFI or by immunodot alone

	IFI+ / Immunodot+ anti-centromere antibodies (n=18)	IFI- / Immunodot+ anti-centromere antibodies (n=43)
IFI :		
Aspect of NAA	Centromeric spotted dog: 15 Centromeric spotted dog: 3	Speckled: 34 Nucleolar speckled: 6 Homogenous: 3
AAN title		
1/160	0	1 (2,3%)
1/320	2 (11,1%)	8 (18,6%)
1/640	3 (16,7%)	9 (20,9%)
1/1280	13 (72,2%)	25 (58,2%)
Immunodot :		
Anti-CENP-B		
+	3 (16,7%)	26 (60,5%)
++	1 (5,5%)	7 (16,3%)
+++	14 (77,8%)	10 (23,2%)
Other special features		
no	8 (44,4%)	20 (46,5%)
one or more	10 (55,6%)	23 (53,5%)
Sm/RNP	1 (5,5%)	4 (9,3%)
Sm	0	1 (2,3%)

SS-A	2 (11,1%)	5 (11,6%)
Ro-52	3 (16,7%)	7 (16,3%)
SS-B	1 (5,5%)	1 (2,3%)
Scl-70	0	2 (4,6%)
PM-Scl 100	1 (5,5%)	1 (2,3%)
PCNA	0	3 (7%)
Nucleosomes	0	3 (7%)
Histones	0	5 (11,6%)
DFS70	2 (11,1%)	5 (11,6%)
Jo-1	1 (5,5%)	1 (2,3%)
AMA-M2	4 (22,2%)	3 (7%)

6. CLINICAL SIGNIFICANCE OF ANTI-CENTROMERE AC POSITIVITY

Clinical information was available for 18 patients. The main reasons given were haematological (33%), articular (27%) and cutaneous (27%) symptoms (Table II). The diagnosis of a connective tissue disease (SSc, SLE, rheumatoid arthritis) was made in 3 patients (Table III).

Table II: Main reasons given for prescribing NAAs

Events	Number (Percentage)
Haematological	6 (33%)
Joint	5 (27%)
Skin	5 (27%)
Dry syndrome	4 (22%)
Lung	3 (16%)
Renal	3 (16%)
Neurological	3 (16%)
Hepatic	3(16%)
Digestive	2 (11%)
Muscular	2 (11%)
Serositis	2 (11%)

Table III: Profile of patients with a diagnosis of connective tissue

disease

Diagnosis	Gender	Age (years)	AAN by IFI on Hep-2 cells	anti-CENP-B by immunodot	Remaining typing by immunodot
ScS	female	20	1/1280 speckled	+++	Nucleosome, Histones, Sm, Sm/RNP, SSA, Ro52, Scl70
LES	female	24	1/1280 speckled	+	Nucleosome, Histones, Sm/RNP
Rheumatoid arthritis	female	29	1/1280 speckled	+	negative

DISCUSSION

In this study, we looked at the positivity of anti-centromeric Ac by IFI and/or immunodot in clinical practice. Our results showed a seroprevalence of 2.1% among requests for serum tests for NAA that had been screened by IFI on Hep-2 cells and typed by immunodot. This frequency is comparable to that reported by Kaaouch H et al (11/821; 1.3%) (7) and Pakunpanya K et al (40/3233; 1.23%) (8). One of the parameters influencing the frequency of detection of autoAbs is the technique used for their detection. In our work, anti-centromere Ac were detected by IFI and/or immunodot. IFI on Hep-2 cells is the gold standard for the detection of NAA (9). Hep-2 cells represent the substrate of choice because their nuclei are large with several nucleoli, allowing better definition of nuclear fluorescence aspects. The smear includes numerous dividing cells with cells at different stages of the cell cycle, which facilitates the detection of NAAs directed against antigenic targets present only at certain phases of the cell cycle (centromere, PCNA, mitotic structures). In addition, the cytoplasm of the cells is abundant, allowing the detection of anti-cytoplasmic Ac (mitochondria, ribosomes, etc.). In addition, the cells are fixed so that the Ag can be preserved in their natural conformation. IFI on Hep-2 cells is therefore a sensitive technique that can detect many auto-Ag at the same time. However, it is an operator-dependent and subjective technique. In order to harmonise the different names and descriptions of the fluorescence aspects observed, a consensus was reached leading to the distinction of 15 aspects of nuclear fluorescence (AC-1-AC- 14 and AC-29), 9 aspects of cytoplasmic fluorescence (AC-15-AC-23) and 5 aspects of mitotic fluorescence (AC-24-AC-28) (10). Each aspect guides towards one or more antigenic targets, but the rule is not absolute (5)(10). It may be observed in isolation or in association with other aspects. Unlike other aspects of nuclear fluorescence, the centromeric aspect (AC-3) on IFI makes it possible to retain the positivity of anti-centromeric Ac without the need for confirmation by a mono-specific technique (10). Immunodot is a sensitive enzyme-linked immunosorbent assay that detects a number of specific autoantibodies in parallel, depending on the number of different antigens deposited on the strip. The kit used in our work can detect 15 other types of autoantibodies in addition to anti-CENP-B antibodies. The antigens used are recombinant (CENP-B, Ro-52, PM-Scl 100, PCNA, DFS70) or purified native (nRNP/Sm, Sm, SS-A, SS-B, Scl-70, Jo-1, native DNA, nucleosomes, histones, ribosomal protein P, AMA- M2) and of human (for recombinant Ag) or animal (for purified Ag) origin. Interpretation is made easy by the appearance of a more or less intense band at the site of the Ag-Ac reaction, and is made more objective by reading the results on a scanner. The result is semi-quantitative, depending on the intensity of the band. Among 3,000 requests for serum testing for NAA that had been screened both by IFI on Hep-2 cells and by immunodot typing, the results of the tests for anti-centromeric Ac (by IFI) and anti-CENP-B (by immunodot) were concordant in 18 cases (positivity of anti-centromeric Ac by the two techniques) and also in 2,937 cases not included in our study (negativity of anti-centromeric Ac by the two techniques), i.e. concordance of the two techniques in 98.5% of cases. The cases of discordance (45 cases, i.e. 1.5%) corresponded in the majority of cases to a negativity of anti-centromeric Ac on IFI and a positivity of anti-CENP-B Ac on immunodot (43 cases/45). On the one hand, this may be explained by the fact that one aspect of fluorescence may mask another; the centromeric appearance could therefore be masked by speckled or homogeneous nuclear fluorescence when the latter is of identical or higher intensity. On the other hand, the presence or absence of centromeric fluorescence on IFI could depend on the titre of anti-centromeric Ac. In fact, anti-CENP-B Ac positivity at 2 or 3 crosses was significantly more frequent in the case of IFI + than in the case of IFI-.The other cases of discordance, less frequent, corresponded to a positivity of anti-centromeric Ac by IFI and a negativity of anti-CENP-B Ac by immunodot (2 cases/45). This situation may be explained by the presence of anti-centromere antibodies with specificity other than centromeric protein B. In fact, CENP-B is the major autoAg recognised by anti-centromere antibodies, which justifies its inclusion in the "classic" panel of Ag tested routinely for AAN typing. However, other auto-Ag can be recognised by anti-centromere Ac, notably CENP-A, CENP-C, CBX5 and MIS12C (11). Kajio et al. have identified other antigenic targets of anti-centromeric Ac such as CENP-HIKM, CENP-TWSX, CENP-OPQUR, CENP-LN, NDC80C, KNL1C and the Astrin-SKAP complex (Figure 13) (12). These authors showed that 14%-23% of patients with SSc, Gougerot-Sjögren's syndrome or primary biliary cirrhosis had autoAb directed against several centromeric antigens but no anti-CENP-B antibodies (12).

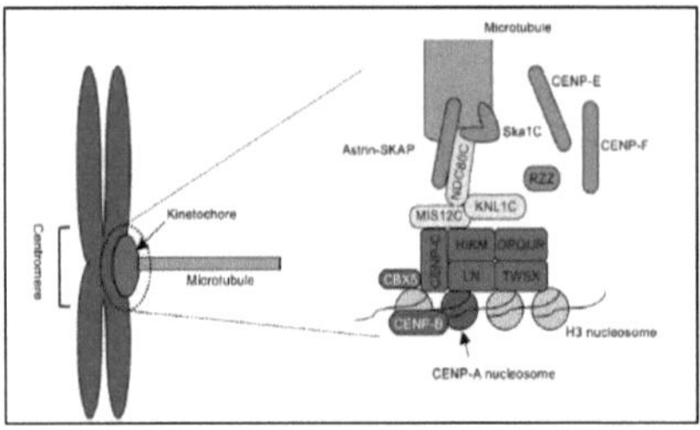

Figure 13: Antigenic targets of anti-centromeric Ac (12)

In addition to IFI and immunodot, other techniques for detecting anti-centromere Ac (anti-CENP-B and/or anti-CENP-A) include western blot (13) and ELISA (Enzyme-Linked Immunosorbent Assay) (14)(15). Several authors have compared these different techniques and the different kits on the market. The degree of agreement varies according to the population tested and the positivity threshold considered (Table IV).

Table IV: Degree of concordance of different techniques for the detection of anti-centromeric Ac according to studies

Study	Techniques /Kits compared	Concordance (Cohen's kappa coefficient)
Bost et al, 2021 (16)	*Anti-CENP-A by immunodot and IFI *Anti-CENP-B by immunodot and IFI	0.47 - 0.91 (depending on patient group) 0.77 - 0.97 (depending on patient group)
Alkema et al, 2021 (15)	*Anti-CENP-A by immunodot: 2 different kits *Anti-CENP-B by immunodot: 2 different kits Anti-CENP-B by immunodot: (1er kit) and anti-CENP-B by ELISA Anti-CENP-B by immunodot: (2ème kit) and anti-CENP-B by ELISA	0.99 0.99 0.96 0.97
Bonroy et al, 2012 (13)	*(Anti-CENP-A or anti-CENP-B by immunodot) and (IFI or western blot) *(Anti-CENP-A and anti-CENP-B by immunodot) and (IFI or western blot)	0.848 0.820
Mahler et al, 2011 (14)	*Anti-CENP-A by ELISA and IFI *Anti-CENP-B by ELISA and IFI *Anti-CENP-A by ELISA and anti-CENP-B by ELISA	0.82 - 0.92 (depending on the positivity threshold) 0.54 - 0.93 (depending on the positivity threshold) 0.55 - 0.91(depending on the positivity threshold)

In addition to the technical aspects, we were interested in the clinical characteristics of patients with positive anti-centromere Ac. Demographically, we noted a clear predominance of women (89%), which is consistent with the results of other studies carried out in

Morocco (81.8%) (7), Thailand (85%) (8), Spain (90.3%) (17) and Japan (94%) (18). Generally speaking, autoimmune diseases are characterised by the fact that they are more frequent in women than in men (19). This may be explained by the action of sex steroid hormones, but also by genetic factors linked to the X chromosome (19)(20). The average age of the cases included in our study was 39.7 years, whereas it was higher in other studies (50 years in the study by Kaaouch H et al (7), 60 years in the study by Tsukamoto M et al (18), 64 years in the study by Alende-Castro V et al (17). It should be noted that in our study, paediatric cases of anti-centromere Ac positivity were noted. Clinically, patients with positive anti-centromere antibodies presented with a variety of symptoms (articular, cutaneous, pulmonary, etc.) and were mainly followed up in the internal medicine department, but also by other medical specialities. Since they were first described in 1980, anti-centromere antibodies have been associated with SSc (21); SSc is a systemic autoimmune disease, belonging to the group of connectivites, with a heterogeneous clinical expression and the possibility of involvement of different organs. The positivity of anti-centromeric Ac is part of the ACR/EULAR classification criteria for SSc (22). These auto-Ac are detected in 20-30% of patients with SSc (23). However, a lower frequency (6.5%) has been reported in a Tunisian population (24). Although their sensitivity is relatively low, their specificity has been reported to be high (>95%). (25). However, some authors have disputed the high specificity of anti-centromere antibodies for SSc (6). The percentage of patients with SSc among those with anti-centromere Ac was 25% for Pakunpanya K et al (8), 36% for Tsukamoto M et al (18), 55.9% for Alende-castro V et al (17), 67% for Roberts-Thomson PJ et al (26) and 70% for Miyawaki S et al (27). The divergence in results between these different studies could be explained by technical variability but also by the population studied, the degree of clinical suspicion motivating the search for these auto-Ac (pre-test probability) and the criteria adopted for the diagnosis of SSc. Our study, although limited by the small number of cases with clinical information available, supports the idea that anti-centromere antibodies are not specific to SSc. Indeed, these antibodies have been reported in other autoimmune diseases including Gougerot-Sjögren's syndrome (28), SLE (29), rheumatoid arthritis (30) and primary biliary cholangitis (31). In addition, genuine overlapping syndromes between SSc and another autoimmune disease may be observed. Serologically, autoAbs associated with different autoimmune diseases may be detected concomitantly in the same patient. In our study, positive anti-CENP-B antibodies were associated with immunodot detection of one or more other antigenic specificities in 31/61 cases, in particular anti-Ro-52, anti-SS-A (associated with Gougerot-Sjögren's syndrome and SLE) and anti-M2 (markers of primary biliary cholangitis). It should be noted that anti-CENP-B and anti-Scl70 were detected concomitantly in two cases, whereas these autoantibodies are reported to be mutually exclusive (32). It has also been reported that anti-centromere antibodies are associated with the limited cutaneous form of SSc, unlike anti-Scl70 antibodies, which are associated with the diffuse cutaneous form of the disease (23). However, the frequency of coincidence of these two types of auto Ac is 0.5 to 0.6% (33)(34). Given that anti-centromeric Ac positivity has been reported in patients with autoimmune diseases (SSc and others) as well as in patients with other conditions, some authors have investigated factors that may determine the clinical relevance of anti-centromeric Ac. Some have shown that higher titres of anti-centromere antibodies (by IFI on Hep-2 cells) were more frequently observed in the autoimmune disease group than in the non-autoimmune disease group, although the difference was not statistically significant (8). Others have noted that the levels of anti-CENP-B Ac by ELISA were significantly higher in the ScS group compared with the other disease groups (27). In the latter study, it was also noted that isolated anti-centromeric Ac positivity was more frequent in patients with ScS and without overlap syndrome, whereas anti-centromeric Ac associated with other NAA specificities were more frequently observed in patients with other diseases or overlap syndromes (27). Tsukamoto M et al have emphasised the importance of the initial presentation in determining the clinical entity (18): for patients diagnosed with an autoimmune disease, Raynaud's phenomenon, sclerodactyly, anaemia and thrombocytopenia were significantly more frequent; IgG, IgA and IgM immunoglobulin levels were significantly higher (18). It should be noted that, in the context of isolated Raynaud's phenomenon, anti-centromere antibodies are predictive of progression to scleroderma. These antibodies are particularly associated with CREST syndrome, a clinical form of limited SSc associating subcutaneous calcinosis, Raynaud's phenomenon, oesophageal dysmotility, sclerodactyly and telangiectasia, where they are found in 57 to 82% of cases (35). Thus, the interpretation of a positive anti-centromeric antibody must take into account the titre of these antibodies, whether they are positive in isolation or in association with other autoantibodies, and above all the clinical context.

Anti-centromere antibodies are part of the NAAs, which are the most frequently prescribed autoantibodies. Anti-centromere antibodies can be detected by IFI on Hep-2 cells (special fluorescence appearance) and by immunodot using the CENP-B protein as the target antigen. Our aim was to assess the concordance of these two immunological techniques and to determine the clinical significance of the positivity of anti-centromere antibodies detected by these techniques. We conducted a retrospective study over a 2-year period (January 2020-December 2021). We listed the requests for serum testing for AAN sent to the Immunology Laboratory of the Habib Bourguiba University Hospital in Sfax that had undergone both IFI screening on Hep-2 cells and immunodot typing. We included patients presenting a positivity of anti-centromeric Ac by IFI and/or immunodot. These two techniques were performed using commercial kits "IIFT: HEp-2" and "Euroline ANA Profile 3 plus DFS70 (IgG)"."respectively. During the study period, 63 cases of positive anti-centromere antibodies (by IFI and/or immunodot) were included, representing 2.1% of NAA tests (IFI+immunodot). The mean age was 39.7 years; the sex ratio was 7 men/56 women. Requests were referred by the internal medicine department in 31.7% of cases. NAA titres ranged from 1/160 (1.5%) to 1/1280 (61.9%). The results of tests for anti-centromere antibodies (by IFI) and anti-CENP-B antibodies (by immunodot) were concordant in 18 cases (anti-centromere antibodies positive by both techniques) but also in 2937 cases not included in our study (anti-centromere antibodies negative by both techniques), i.e. concordance between the two techniques in 98.5% of cases. The discrepancies were 2 cases IFI+/immunodot- and 43 cases IFI-. /immunodot+. In these 43 cases, the appearance of NAA on IFI was mottled (34 cases), Nucleolar speckle (6 cases) or homogeneous (3 cases); anti-CENP-B antibodies were weakly positive (26 cases), positive (7 cases) or strongly positive (10 cases). No specificity other than CENP-B was detected by immunodot in 30 cases. In the other cases, the most frequently associated specificities were anti-Ro52(n=10), anti-SSA(n=7), anti-M2(n=7) and anti-DFS70(n=7). Anti-CENP-B Ac positivity at 2 or 3 crosses was significantly more frequent in the case of IFI + (15/18; 83.3%) than in the case of IFI- (17/43; 39.5%) (p=0.002). There was no significant difference between the two groups in the association of anti-CENP-B antibodies with other antigenic specificities on immunodot (55.6% versus 53.5%; p=0.883). Of the 18 patients for whom clinical information was available, 3 were diagnosed as having a connective tissue disease (SSc, SLE, rheumatoid arthritis). Our results are consistent with the literature in terms of the frequency of routine detection of anti-centromeric Ac, the predominance of females in patients who are seropositive for these Ac, the heterogeneity of the clinical signs encountered and the good agreement between the IFI and immunodot techniques. IFI, which is indicated as the first-line method for screening for NAA, has a number of advantages, but also certain limitations (operator-dependence; one aspect of fluorescence may be masked by another). Immunoblot is useful for confirming antigen specificity, within the limits of the antigens used (in our study, only anti-CENP-B antibodies were detected). Other techniques, such as ELISA, can be used routinely. With regard to their clinical relevance, anti-centromere antibodies have been reported to be specific markers of SSc; however, their specificity has been disputed by some authors. Our study, although limited by the small number of cases with clinical information available, supports the idea that anti-centromeric Ac are not specific to SSc and can be detected in other autoimmune diseases (including SLE, The clinical relevance of anti-centromeric antibody positivity depends on the titre of the anti-centromeric antibody, whether it is isolated or associated with other autoantibodies and, above all, on the clinical context. The clinical relevance of a positive anti-centromere antibody depends on the titre of these antibodies, whether they are positive in isolation or in association with other autoantibodies, and above all on the clinical context. In this work, we reiterate the importance of mastering immunological techniques. Knowledge of the advantages and limitations of each technique helps to explain the discrepancies encountered in practice. Comparison of the results with clinical information is necessary for better interpretation, underlining the importance of clinician-biologist dialogue.

REFERENCES

1. Pasquali J, Goetz J. What should be done in the presence of antinuclear antibodies in adults? In: Lupus erythematosus. 2013. p. 103.

2. Masson C, Bouvard B, Houitte R, Petit A, Manach L, Hoppe E, et al. Clinical interest of antinuclear antibodies: The expectation of the rheumatologist during systemic diseases. Rev Francoph des Lab. 2006;(384):71-6.

3. Bossuyt X, Meroni PL. Understanding and interpreting antinuclear antibody tests in systemic rheumatic diseases. Nat Rev Rheumatol [Internet]. 2020; Available from: http://dx.doi.org/10.1038/s41584-020- 00522-w

4. Goulvestre C. Antinuclear antibodies. Presse Med. 2006;35(2):287-95.

5. Lassoued K, Coppo P, Gouilleux-Gruart V. Place of antinuclear antibodies in clinical practice? Réanimation [Internet]. 2005;14(7):651-6. Available from:
http://linkinghub.elsevier.com/retrieve/pii/S1624069305001829

6. Roberts-Thomson P. Specifcity of anti-centromere antibodies for scleroderma. Rheumatol Int. 2007;28:197-8.

7. Kaaouch H, Bhallil O. Anti-centromere antibodies and associated autoimmune diseases. Qatar Med J. 2023;2023(2):1-2.

8. Pakunpanya K, Verasertniyom O, Vanichapuntu M, Pisitkun P, Totemchokchyakarn K, Nantiruj K, et al. Incidence and clinical correlation of anticentromere antibody in Thai patients. Clin Rheumatol. 2006;25(3):325-8.

9. Meroni PL, Schur PH. ANA screening: an old test with new recommendations. Ann Rheum Dis [Internet]. 2010;69(8):1420-2. Available from: http://www.ncbi.nlm.nih.gov/pubmed/20511607

10. Damoiseaux J, Eduardo L, Andrade C, Carballo OG, Conrad K, Luiz P, et al. Clinical relevance of HEp-2 indirect immunofluorescent patterns: the International Consensus on ANA patterns (ICAP) perspective. 2019;879- 89.

11. Stochmal A, Czuwara J, Trojanowska M, Rudnicka L. Antinuclear Antibodies in Systemic Sclerosis: an Update. Clin Rev Allergy Immunol. 2019;2-8.

12. Kajio N, Takeshita M, Suzuki K, Kaneda Y, Yamane H, Ikeura K,

et al. Anti-centromere antibodies target centromere - kinetochore macrocomplex: a comprehensive autoantigen profiling. 2021;651-9.

13. Bonroy C, Praet J Van, Smith V, Steendam K Van, Mimori T, Deschepper E, et al. Optimization and diagnostic performance of a single multiparameter lineblot in the serological workup of systemic sclerosis. J Immunol Methods [Internet]. 2012;379(1-2):53-60. Available from: http://dx.doi.org/10.1016/j.jim.2012.03.001

14. Mahler M, You D, Baron M, Taillefer SS, Hudson M, Scleroderma C, et al. Anti-centromere antibodies in a large cohort of systemic sclerosis patients: Comparison between immuno fl uorescence , CENP-A and CENP-B ELISA. Clin Chim Acta [Internet]. 2011;412(21–22):1937–43. Available from: http://dx.doi.org/10.1016/j.cca.2011.06.041

15. Alkema W, Koenen H, Kersten BE, Kaffa C, Dinnissen JWB, Damoiseaux JGMC, et al. Autoantibody profiles in systemic sclerosis; a comparison of diagnostic tests. Autoimmunity [Internet]. 2021;54(3):148-55. Available from: https://doi.org/10.1080/08916934.2021.1907842

16. Bost C, Fortenfant F, Antoine B, Pugnet G, Renaudineau Y. Combining multi-antigenic immunodot with indirect

immunofluorescence on HEp-2 cells improves the diagnosis of systemic sclerosis. Clin Immunol. 2021;229(June).

17. Alende-castro V, Vázquez-triñanes C, Rodríguez-fernández S. Significance of Anti-Centromere Antibodies. Int J Med Sci Heal Res. 2020;4(03):147-58.

18. Tsukamoto M. Initial presentation determines clinical entity in patients with anti - centromere antibody positivity. Int J Rheum Dis. 2018;(August):1-5.

19. Le Guern V. Sex hormones and autoimmunity. La Press Médicale Form[Internet].   2020;1(1):36-41.Available from: https://doi.org/10.1016/j.lpmfor.2020.03.019

20. Miquel CH, Youness A, Guéry JC. Female predominance of autoimmune diseases: Do lymphocytes have a sex? Rev du Rhum Monogr. 2021;88(1):3-7.

21. Moroi Y, Peebles C, Fritzler MJ, Steigerwald J, Tan EM. Autoantibody to centromere (kinetochore) in scleroderma sera. Proc Natl Acad Sci U S A. 1980;77(3 I):1627-31.

22. Van Den Hoogen F, Khanna D, Fransen J, Johnson SR, Baron M, Tyndall A, et al. 2013 classification criteria for systemic sclerosis: An

American college of rheumatology/European league against rheumatism collaborative initiative. Ann Rheum Dis. 2013;72(11):1747-55.

23. Hamaguchi Y, Takehara K. Anti-nuclear autoantibodies in systemic sclerosis: News and perspectives. J Scleroderma Relat Disord. 2018;3(3):201-13.

24. Salah R Ben, Frikha F, Hachicha H, Chabchoub I, Dammak C, Snoussi M, et al. Clinical and serological profile of systemic sclerosis in Tunisia: A retrospective observational study. Presse Med [Internet]. 2019; Available from: https://doi.org/10.1016/j.lpm.2019.07.037

25. Damoiseaux J, Potjewijd J, Smeets RL, Bonroy C. Autoantibodies in the disease criteria for systemic sclerosis: The need for specification for optimal application. J Transl Autoimmun [Internet]. 2022;5(January):100141. Available from: https://doi.org/10.1016/j.jtauto.2022.100141

26. Roberts-Thomson PJ, Nikoloutsopoulos T, Cox S, Walker JG, Gordon TP. Antinuclear antibody testing in a regional immunopathology laboratory. Immunol Cell Biol. 2003;81(5):409-12.

27. Miyawaki S, Asanuma H, Nishiyama S, Yoshinaga Y. Clinical

and serological heterogeneity in patients with anticentromere antibodies. J Rheumatol. 2005;32(8):1488-94.

28. Bournia VK, Diamanti KD, Vlachoyiannopoulos PG. Anticentromere antibody positive Sjögren's Syndrome: a retrospective descriptive analysis. Arthritis Res Ther. 2010;12.

29. Respaldiza N, Wichmann I, Ocan C. Anti-centromere antibodies in patients with systemic lupus erythematosus. Scand J Rheumatol. 2006;35:290-4.

30. Kuramoto N, Ohmura K, Ikari K, Yano K, Furu M, Yamakawa N, et al. Anti-centromere antibody exhibits specific distribution levels among anti- nuclear antibodies and may characterize a distinct subset in rheumatoid arthritis. Sci Rep. 2017;7(1):1-8.

31. Liberal R, Grant CR, Sakkas L, Bizzaro N, Bogdanos DP. Diagnostic and clinical significance of anti-centromere antibodies in primary biliary cirrhosis. Clin Res Hepatol Gastroenterol [Internet]. 2013; Available from: http://dx.doi.org/10.1016/j.clinre.2013.04.005

32. Kikuchi M, Inagaki T. Bibliographical Study of the Concurrent Existence of Anticentromere and Antitopoisomerase I Antibodies. Clin Rheumatol. 2000;19:435-41.

33. Dick T, Mierau R, Bartz-Bazzanella P, Alavi M, Stoyanova-Scholz M, Kindler J, et al. Coexistence of antitopoisomerase I and anticentromere antibodies in patients with systemic sclerosis. Ann Rheum Dis. 2002;61:121-7.

34. Heijnen IAFM, Foocharoen C, Bannert B, Carreira PE, Caporali R, Smith V, et al. Clinical significance of coexisting antitopoisomerase I and anticentromere antibodies in patients with systemic sclerosis: an EUSTAR group-based study. Exp Rheumatol. 2013;(11):96-102.

35. Admou B, Essaadouni L, Amal S, Arji N, Chabaa L, Aouad R El. Autoantibodies in systemic scleroderma: clinical interest and diagnostic approach. Ann Biol Clin. 2009;67(3):273-81.

SUMMARY

Anti-centromere antibodies (ACA) can be detected by indirect immunofluorescence (IFI) or immunodot (ID). Our aim was to assess IFI/ID concordance and the clinical relevance of ACA. Patients with positive ACA by IFI on Hep-2 cells and/or by ID were included. ACAs were detected in 63 cases (2.1% of IFI+ID tests). ACA were positive by IFI and ID in 18 cases. The discrepancies were 2 cases of ACA-IFI+/ACA-ID- and 43 cases of ACA-IFI-/ACA-ID+. For these 43 cases, the appearance of nuclear fluorescence (IFI) was speckled (34), nucleolar speckled (6) or homogeneous (3); ACA (ID) was weakly positive (26), positive (7) or strongly positive (10). The diagnosis of a connective tissue disease (systemic scleroderma, systemic lupus erythematosus, rheumatoid arthritis) was made in 3 patients out of 18 with clinical information. IFI on Hep-2 cells has certain limitations (operator-dependent; one aspect of fluorescence may be masked by another). ID is useful for confirming antigen specificity, within the limits of the antigens used. Interpretation of these tests depends on the clinical context.

I want morebooks!

Buy your books fast and straightforward online - at one of world's fastest growing online book stores! Environmentally sound due to Print-on-Demand technologies.

Buy your books online at
www.morebooks.shop

Kaufen Sie Ihre Bücher schnell und unkompliziert online – auf einer der am schnellsten wachsenden Buchhandelsplattformen weltweit! Dank Print-On-Demand umwelt- und ressourcenschonend produziert.

Bücher schneller online kaufen
www.morebooks.shop

Printed by Books on Demand GmbH, Norderstedt / Germany